Keeping Mother, Baby and Child Safe from Toxic Chemicals

Professor Norman Ratcliffe

A catalogue record for this book is available from the British Library

ISBN: 978-1-907962-68-4

Published by Cranmore Publications

www.cranmorepublications.co.uk

All Parents Should Read This Book Because Babies And Children Are Potentially More Vulnerable Than Adults To Toxic Chemicals

The Author

Professor Norman Ratcliffe is a founder member of a team that recently discovered a new antibiotic potentially capable of curing MRSA and *Clostridium difficile*. This work was presented to Prince Phillip at St. James's Palace, London and was the subject of major media attention in the UK on ITV News and in many leading newspapers around the world, including the Wall Street Journal. He is a Fellow of the Royal Society of Medicine and has previously run a "Health Alert" blood-testing company. He has published over 200 books and research papers on immunology, cancer invasion, influenza, tropical diseases and MRSA. He played squash for Wales, ran the London Marathon at the age of 50 and works-out regularly in the gym.

Professor Ratcliffe retired recently after 25 years as a University Research Professor. He decided to finally complete his comprehensive book on health *"It's Your Life: End the Confusion From Inconsistent Health Advice"* after 5 years work in order to help the many people who are confused about health and fitness issues and who have constantly been seeking his advice.

Contents

Contents

Some Basic Facts

Our bodies are constantly exposed to numerous potentially toxic chemicals from pesticides and additives in our food and drink, as well as environmental contaminants from cars, cosmetics and furnishings. These chemicals accumulate in the body to form the "Body Burden", a chemical "Cocktail" with unknown health consequences.

THERE IS NO ESCAPE FROM THESE CHEMICALS!

Except to go and live in a deep cave with no modern appliances or furnishings, processed food, drink or consumer products!!

It is most important to understand that unborn babies in the womb, as well as babies and rapidly developing children, are likely to be more sensitive to chemical contaminants than are adults.

For example, **thyroid hormones (eg. thyroxine)** play key roles in babies' brain development, and pollutants, such as PCBs (previously released into the environment from numerous industrial processes), may decrease thyroxine levels and disrupt normal brain development.

These contaminants can be passed on to the foetus and baby via the mother's milk at all stages of development. They can also adversely affect the foetus well below levels described as "safe" in the adult woman.

There are particular worries about the possible effects of **pesticide residues in food** on the delicate and rapidly growing bodies of **babies and young children.** In 1996, the USA National Academy of Science confirmed that any health risks of such chemicals would be magnified in the young.

Introduction

This book will identify sources of the main toxic chemicals taken into our bodies and advise **women planning a pregnancy, pregnant, or with young children,** how to reduce levels of these harmful substances.

Read the lists of toxic substances and advice on how to avoid these. Then begin by selecting simple changes to your lifestyle which will reduce contact with these toxins. Nobody has time or money to follow all the advice given. For example, you could

1. Replace perfumed soap with Simple Soap and reduce the use of cosmetics while pregnant.

2. Make sure that you thoroughly wash fruit and vegetables.

3. Always wear rubber gloves for cleaning.

4. Make sure that the processed meals that you buy in the Supermarket do not contain harmful colourants or other additives etc.

5. Get rid of the old and flaky, non-stick pan.

6. Throw away all that PVC plastic food wrapping.

7. Start to wean yourself off all those zero cokes!

8. Throw away the tins of baby food and replace with jars etc.

Gradually, change things so that you reduce your body burden of chemicals and make it a safer home for your baby to develop in.

The outline of the rest of the book is as follows:

Chapter 1 identifies sources of toxic chemicals in the body.

Chapter 2 explores the existence of pesticides in food.

Chapter 3 looks at additives in food.

Chapter 4 identifies environmental contaminants.

Chapter 5 deals with detox programmes prior to pregnancy.

Chapter 6 explores which organic foods are most beneficial.

Chapter 1

Sources of Toxic Chemicals in the Body

These come from:

- **Pesticides** used to control insects, slugs, rats and microbes on fruit, vegetables, cereals and animals.

- **Additives** to food and drink including colourants, preservatives, sweeteners and flavourings.

- **Environmental contaminants** passing through the skin and into the lungs from furnishings, household products, cosmetics, cars, plastics, cigarettes, industries etc. (see figure below).

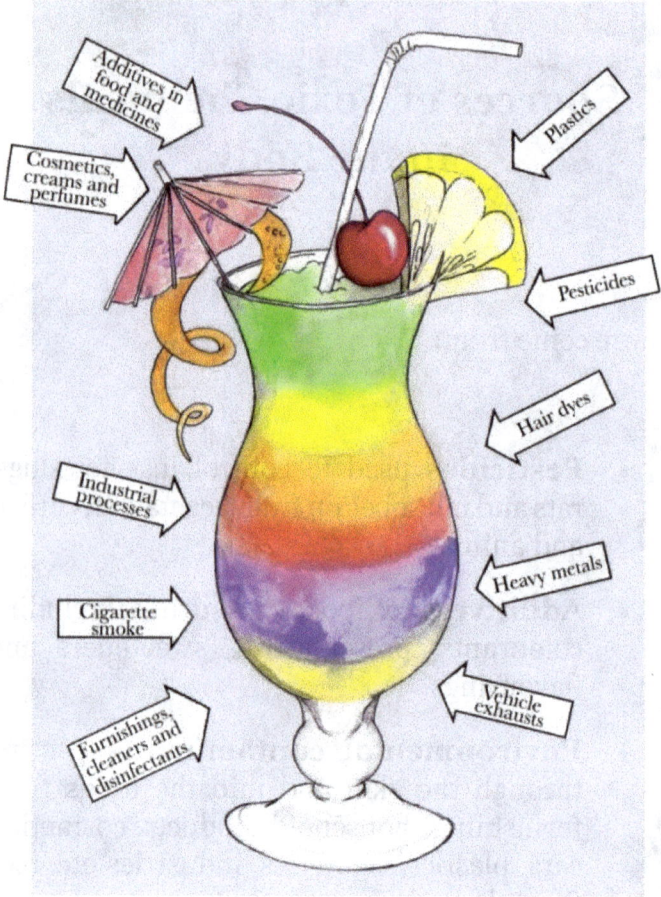

Additives in food and medicines

Cosmetics, creams and perfumes

Plastics

Pesticides

Hair dyes

Industrial processes

Heavy metals

Cigarette smoke

Vehicle exhausts

Furnishings, cleaners and disinfectants

In 2007, **the Environmental Working Group, Human Toxome Project,** tested for 532 different contaminant chemicals in the blood, urine, or breast milk of 174 American people from babies in the uterus or newborn, teens, adults and seniors (over 65 yr). Chemicals detected included:

1. Pesticides used to kill insects, slugs etc.

2. Parabens and PAHs, used in cosmetics, creams, toothpaste etc.

3. Bisphenol, (BPA) found in plastics including babies bottles.

4. PBDE, used as flame retardants in furnishings.

5. PCBs, banned but previously released from numerous industrial processes.

6. Dioxins, unintentionally released by industrial processes.

7. Many others including PFCs used in non-stick frying pans and heavy metals such as lead and mercury.

The presence of dioxins, mercury, PFCs and PCBs were detected in newborn babies and these are of particular concern as they may be associated with hormonal disturbances, immune impairment, allergies, cancer, learning difficulties and behavioural problems.

Children of 2 years and younger have 10 times the risk of adults after exposure to toxins since they breathe more quickly, have thinner skin and eat and drink relatively more than adults.

(www.cleanhouston.org/health/health_effects/health7.htm).

Chapter 2

Pesticides in Your Food

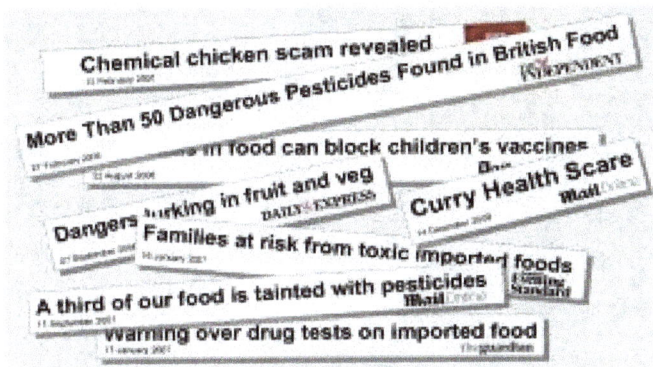

Chemical chicken scam revealed

More Than 50 Dangerous Pesticides Found in British Food

...in food can block children's vaccines

Dangers lurking in fruit and veg

Curry Health Scare

Families at risk from toxic imported foods

A third of our food is tainted with pesticides

Warning over drug tests on imported food

*It's Easy to Clean up Your Act
and Avoid Chemical Pesticides*

Sources of pesticides are illustrated in the figure below:

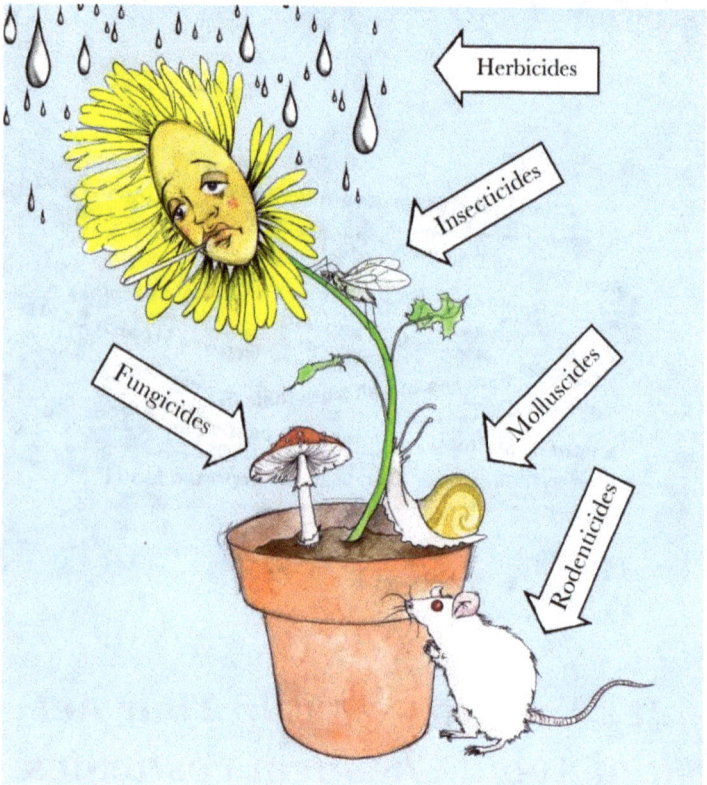

Pesticides are groups of chemicals used to control pests attacking our food crops. The results of the use of these pesticides are not entirely bad and it has been estimated that without the use of artificial nitrogen fertilizers, the World's available fertile land could only support about 60% of the present population. However, with the incidences of some diseases, such as cancers and allergies, increasing at alarming rates, the environmental damage caused by the use of chemical pesticides has to be rectified. With this aim as a priority, scientists are developing integrated pest management (IPM) programmes, minimising the use of chemicals and reducing the impact on natural animal and plant populations and on humans. These pests and the pesticides used to combat them include:

i. **insects** - insecticide chemicals

ii. **fungal diseases** - fungicides

iii. **weeds** – herbicides

iv. **slugs and snails** – molluscicides

v. **rats and mice** – rodenticides

After the crop has been harvested, very small amounts of these chemicals may remain in the food as **chemical residues** that are consumed by both humans and farmed animals. Many of these chemicals are not rapidly eliminated from the body as they dissolve in fatty tissues and then **accumulate in the body** over some considerable time. Due to their persistence in the environment they are often termed **"Persistent Organic Pollutants or POPS".**

Despite the banning of POPS in many countries (Stockholm Convention on POPs, 2001), they are still being found in human tissues. This is due to **bioconcentration** in food chains, through long-term environmental contamination of soils, water and air by pesticides. As humans are at the top of many food chains, we may ingest high concentrations of these POPS from fish, meat etc.

These residual POPs may cause breast and other cancers, suppress the immune system, disturb hormonal functions and pollute human breast milk with unknown short and long-term consequences for babies and infants. **Our children and unborn babies are particularly vulnerable to pesticides due to their delicate and rapidly growing organs and tissues.**

In 1996, the USA National Academy of Science confirmed that any health risks of such chemicals would be magnified in the young.

More recently developed and widely used replacement pesticides for POPS include the **organophosphates and carbamates**. These are less environmentally persistent than POPS but are more toxic to humans. They too can induce hormonal disturbances, as well as nervous, behavioural, psychiatric and muscular disorders such as those experienced by some Gulf War Veterans.

Pesticide residues in fruit, vegetables and other foods are monitored in the UK by the **Pesticides Residues Monitoring Programme (PRiF).** In the February 2012 report from PRiF, 889 samples of 23 different foods had been surveyed and **26 samples were found to contain residues above the maximum permitted levels** (http://www.pesticides.gov.uk/Resources/CRD/PRiF/Q2_2011_report.pdf).

Risk assessments were carried out by PRiF and effects on health were described as "unlikely" or "not expected" for single pesticides or "from exposures to mixtures "likely to be small". For mixtures of contaminant pesticides PRiF stated there are "-- areas of uncertainty in the risk assessment process and made recommendations for further work".

The public, understandably, still has deep concerns about food safety, and, therefore, useful Tables have been produced of the pesticide contamination rates of fruit and vegetables based on the PRiF's own monitoring data (you can find the Tables in *"It's Your Life: End the Confusion from Inconsistent Health Advice"* by Professor Dr. Norman Ratcliffe).

A summary of contamination rates of fruit, vegetables, cereals, fish, meats and dairy products are given below.

Use this information to clean up your act and reduce the amount of pesticides in your food.

i. FRUITS

Lowest rates!

- All fruits with peel or skins, such as citrus fruits, bananas, mangoes, pineapples, melons, apples etc, probably have the lowest rates of pesticide contamination **AFTER THE PEEL/SKIN IS REMOVED**.

Highest rates!

- In contrast, many "soft" fruits, such grapes, strawberries, apricots, raspberries, cherries, plums, and peaches as well as unpeeled apples and pears, have the highest rates of pesticide contamination. Most of these are not peeled before eating and pesticides probably bind to the skins.

- Organic fruit is expensive so give priority to buying organic soft fruit or apples and pears if you eat the skin. These are not always available and are expensive so wash non-organic soft fruit (uncut to avoid loss of vitamins) for 30sec to 5 min or peel the apples.

ii. VEGETABLES/SALADS

Lowest rates!

- All vegetables that are peeled/podded before eating, such as peas, peeled potatoes, tinned/peeled tomatoes, mushrooms, turnips, parsnips, onions, leeks, carrots, swedes, corn, broad beans, baked beans, etc, probably have the lowest rates of pesticide contamination.

These have low rates too!

- Some items with no peel also have low contamination rates, including cabbage, okra, fennel, broccoli, asparagus, cauliflower, and marrows.

Highest rates!

- In contrast, items with no peel or unpeeled before use, such as bagged salad, fresh lettuce, celery, aubergines, peas with edible pods, green beans, salad onions as well as unpeeled potatoes, fresh tomatoes, cucumber, peppers and courgettes, have the highest rates of pesticide contamination.

- Organic vegetables* are more expensive and often unavailable so always wash and/or trim all non-organic and unpeeled vegetables and salads. Always buy organic potatoes which are relatively cheap and available.

- Beware bought chips and crisps since these may have high pesticide rates from contaminated cooking oil. Make your own chips if you must.

* See *Chapter 6* for further advice on organic food.

iii. **Cereals and flour products**

- Cereals and cereal products, such as wheat flour, rye, bran, oats, rice, bread, cereal bars, and breakfast cereals, have high rates of pesticide contamination.

- Try and buy **organic wholemeal bread, porridge, flour and cereal bars.**

iv. **Fish**

- Many fish contain traces of mercury so that children under 16 as well as pregnant and women planning pregnancy should avoid shark, marlin and swordfish and limit themselves to two fresh tuna steaks per week. Tinned tuna is safer. **Mercury may affect the nervous system of the unborn baby and young children**. Cod, haddock and plaice are fine.

v. **Meats and dairy products**

- Although meats and dairy produce have low rates of pesticide contamination, antibiotics are sometimes added to animal feeds outside the EU* and these may contaminate imported foods. It is better to buy UK* produce if you are certain of the origin. Organic meat and dairy produce is readily available but expensive. Marks and Spencer* and Sainsbury's specialise in quality meats and poultry and sometimes are as cheap as other higher volume supermarkets.

- Trim off excess fat from meat as this is where contaminants may be concentrated.

*In countries outside Europe, it will be necessary to enquire whether antibiotics are used in animal feeds to boost growth.

NB. Note that most organic samples of foodstuffs are free of pesticides but are not regularly available or are too expensive at supermarkets in which case concentrate on trying to buy organic potatoes, bread, cereals (and cereal products) and porridge most of which are usually present on the shelves. Special markets often cater for organic produce in many countries.

Chapter 3

Additives in Your Food

Thousands of foods and drinks contain additives so it is virtually impossible to entirely eliminate all of these from your diet.

Many of these additives are not essential but purely added to improve colour or taste of food which has been lost during processing.

Because an additive has been deemed "safe" does not mean that it is good for you.Many **highly processed foods** with lots of colourants, flavourings and sweeteners contain very few ingredients of any value to the body.

Understand that safe levels of additives (the ADI or acceptable daily intakes) are calculated for each additive when used **on its own** but mixtures of additives, found in many products, may interact and **greatly magnify their toxicity/side effects** ("The Cocktail Effect").

In Europe*, a large number of additives are given E numbers and those of concern generally fall into the following groups:

- Food colourants are usually E100 to E180.

- Preservatives are mainly E200 to E285.

- Antioxidants are mainly E300 to E321.

- Many artificial sweeteners are included in E420, E421 and E950-E968.

- Artificial flavour enhancers e.g. E621.

Details of all the food and drink additives are given in *"It's Your Life: End the Confusion from Inconsistent Health advice"*, by Professor Dr. Norman Ratcliffe.

*Again, outside Europe different ways of identifying food additives may exist but the common names of the chemicals may be given and used to identify the worst culprits (see below).

Note that many food additives are naturally-occurring substances and harmless, such as the colourants, paprika (E160c), lycopene (E160d) and curcumin (E100). Some food additives, such as salt (within the 6gm limit per day) and vinegar (contains acetic acid, E260), have even been used for hundreds of years with no adverse effects reported. However, **concerns have arisen as to the wisdom of adding so many chemicals to our food some of which have proven to be toxic.**

Action to reduce food additives!

i. Food colourants

- These have been shown to induce allergies and cause behavioural problems in children and aggravate allergies in adults. They may also be cancer-forming.

- Azo dyes (incude E102- E129) are synthetic colourants extracted from crude oil and have been used in many foods and medicines but should have been withdrawn voluntarily by food manufacturers in the UK. Some, such as **Tartrazine** (E102), are banned in other countries such as Scandinavia, Germany, USA and Japan. Also, included are **Quinoline Yellow** (E104), **Amaranth red** (E123), **Sunset Yellow** (E110), **Carmoisine** (E122), **Ponceau 4R** also called **Cochineal Red A** (E124), and **Allura Red** (E129).

- Read the food labels and remove from diet/medicines of babies, children and adults.

ii. Food preservatives

- Children are particularly vulnerable to effects of preservatives due to excessive intakes in soft drinks and sweets. Limit intake or dilute juices containing these and provide water.

- **Benzoates** (E210-E214) are widely used in soft drinks, cakes, jam, pickles, sweets, crisps, child medicines, cosmetics, beauty products etc. Safety is controversial and may aggravate allergies and asthma.

- **Sulphites** (metabisulphites etc) (E220-E228) seem to be everywhere including soft drinks, fruit juices, dried fruit, beer, wine, medicine, cosmetics, beauty products etc. **These chemicals are among the top 10**

substances causing allergies including sneezing and breathing problems.

- Again, carefully read labels and remove from diet and from use.

iii. Artificial sweeteners

- A huge range of foods and drinks contain artificial sweeteners so they are difficult to avoid.

- Many foods containing sweeteners are highly processed and of low nutritional value i.e. junk foods such as soft drinks, cakes and sweets.

- The 3 artificial sweeteners of concern are **Aspartame** (E951, banned in Philippines), **Cyclamates** (E952, banned in USA) and **Saccharin** (E953, banned in Canada) have been linked to cancer but evidence is controversial. Young children's drinks containing aspartame should be avoided.

- Of concern is the fact that children and teenagers consume particularly high levels of artificially sweetened junk foods and therefore ingest high doses of these additives.

- Food, drink and medicines containing artificial sweeteners are not recommended for babies (and may actually be banned in baby foods) and very young children (less than 3 yr) as their rapidly developing bodies are particularly vulnerable to any bad effects.

- Most **"diet", "light" or "zero" products contain artificial sweeteners**. The public is being exposed to a high-powered advertising campaign trying to promote these **chemical mixes** to gullible children and adults.

- It is particularly important NOT to expose babies and young children to these artificial sweeteners in any form but particularly "Zero" drink products as they will continue to use these products as they grow older. This may then lead to obesity.

iv. Artificial flavour enhancers

- Thousands of flavourings are used in food and many have not been tested for safety.

- **Monosodium glutamate (MSG)** (E621) is one flavour enhancer widely used and of safety concern for babies and young children. It is found in sausages, pies, sauces, soups, crisps, stock cubes, noodles, Chinese food, tin foods, fast foods, ready meals and some cheeses.

- **MSG** may well be able to cross into the developing brains of newborns and young children and possibly cause damage. Many food producers no longer add MSG to their baby foods.

Chapter 4

Environmental Contaminants

We are constantly in contact with these contaminants in our daily lives in furnishings, household products, cosmetics, cars, plastics, cigarettes, industries etc. These substances are taken in through the skin, lungs, food and drink. Complete avoidance is impossible but action can be taken to reduce contact and protect babies and children.

Action to reduce environmental contaminants!

i. **Plastics -** Take care with these.

- Get rid of any polycarbonate baby bottles (usually marked with no.7 in triangle on bottom of bottles) as they may contain

Bisphenol A (BPA). This is used in making polycarbonate plastics for babies bottles etc and also for lining cans from which it may leach into food and drink. BPA is a hormone disruptor in animal studies. USA and Canada have banned sales of these bottles.

- Never heat or microwave or wash in a dishwasher any plastic baby bottles. Use glass containers to heat baby food.

- Avoid canned food for babies and limit use in adults.

- Throw (recycle!) your plastic water bottles and do not reuse repeatedly.

- Do not store food for long periods in PVC wrap but use cellulose bags. Scrape off a thin layer of cheese stored in PVC before eating.

ii. Health and beauty products

There are **5 chemicals** in these products which are potentially harmful to babies and children and should be reduced.

- **Sodium lauryl sulphate (SLS or SDS)** found in many liquid soaps (washing up liquids are particularly bad due to frequency of use) and shampoos, and causing dermatitis or allergic rhinitis (runny nose and sneezing).

- The **phthalates, such as DBP and DEHP,** and **parabens** (methyl- ,ethyl-, propyl- and butyl –parabens), which are hormone disruptors.

- The **polyethylene glycols (PEGs)** and **propylene glycol** that are irritants or contain carcinogens.

- As an example of avoidance, buy "Simple Soap" without any of these 5 chemicals and also without fragrances which can also be a problem.

iii. Chemicals/irritants in the house

- It is very difficult to avoid these as the average house has many sources of contaminant chemicals in every day products.

- Buy more natural or environmentally safe cleaning agents, for example based on vinegar, available for the toilet, washing up and for cleaning surfaces. Avoid "antibacterial" products (creating superbugs?) and always wear gloves.

- Do not use insect sprays indoors.

- Use natural air fresheners like bunches of lavender in draws and cupboards.

- Avoid scented detergents/conditioners in the laundry and buy those sold specially for babies.

- Use shoe cleaners outside with gloves on.

- Furnishings, carpets and mattresses will be treated with chemicals so try and find out what these are and decide if you want these in your home. Keep the rooms well aerated, if possible, and leave plastic covers in place for a while. In particular, avoid flame retardants (PBDE) which are now banned in the Europe.

- Store paints and cleaners well away from main living areas.

- If you are highly allergic, use house dust mite proof covers on bed linen.

- Vacuum clean regularly to reduce dust laden with household chemicals. Do not forget to vacuum the curtains.

- Use glass, stainless steel or cast iron cooking vessels rather than non-stick and Teflon-coated.

- Avoid dry cleaning clothes due to toxic solvents used.

- **AND, OF COURSE, DO NOT SMOKE OR DRINK!**

Chapter 5

Detox Programmes Prior To Pregnancy

The Body Burden of chemicals will continue to accumulate unless at least some of the above steps are followed to reduce the so called **"Toxic Load".** If you are planning a pregnancy, it makes sense to prevent further build up of contaminants which will subsequently be released into the body over many years. These toxins could potentially harm the unborn baby or be transferred to the baby through the mother's milk.

The value and safety, however, of such detox programmes are the subject of much debate. A successful detoxification programme was, however, undertaken with rescuers at the collapse of the World Trade Centre and some significant reductions of pollutants were recorded which resulted in adverse health symptoms returning to normal.

The best approach is to reduce your Body Burden, as outlined above, making sure that you adopt a balanced diet to control your weight, and participate in a regular exercise programme to prepare for your future pregnancy.

For those who are particularly concerned here are some additional references for detox methods, food and cosmetic additives, and toxins in the environment:

Wisner and colleagues, Treatment of children with detoxification method developed by Hubbard, Proc. Amer. Public Health Assoc. National Conference, 1995.

Rapp, D.J. "Our Toxic World A Wakeup Call", 2004, see:

www.drrapp.com/publications.htm

"Child-specific Exposure Factors Handbook", 2008, USA. Environmental Protection Agency, Washington, USA.

"What's Really in Your Basket? An Easy Guide to Food additives and Cosmetics Ingredients, by Bill Statham, Summersdale Pub. 2007.

Richardson, A. "They Are What You Feed Them", 2006, see:

www.fabresearch.org/view_item.aspx?item_id=960

Chapter 6

Organic Food Is Beneficial So Learn What to Buy

Do we need to buy just organic fruit and vegetables to avoid pesticides?

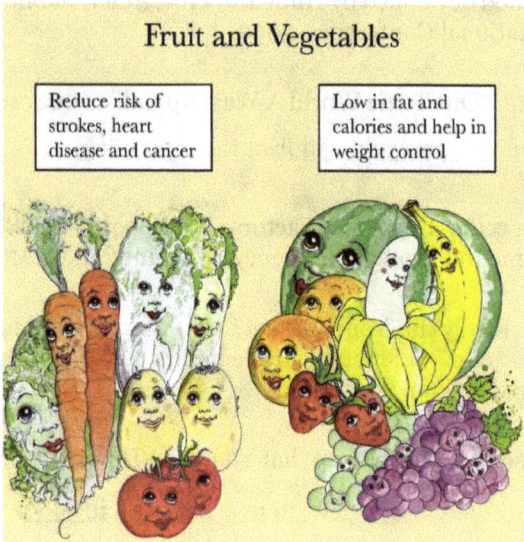

Fruit and Vegetables

Reduce risk of strokes, heart disease and cancer

Low in fat and calories and help in weight control

- **The Conventional Farming Versus Organic Farming** debate still rages. There is an excellent review in "Time" magazine in 2010.

- Most of our food crops are grown by intensive farming with the aid of synthetic fertilizers. According to the **Food Standards Agency** (FSA), the **nutritional content of these foods is no less** than crops grown organically (*www.food.gov.uk/foodindustry*). The FSA is a Government agency set up to protect public health in relation to food.

- **The FSA's conclusion contrasts to the view of the Soil Association** (SA), a charitable organization independent of Government and concerned with promoting organic food and farming, which concluded that organic food has higher levels of vitamin C, as well as of other antioxidants and essential minerals (**Organic Food: Facts and Figures, 2006,** *www.soilassociation.org*).

- The SA's conclusion was based on a review of 41 studies from around the World and included evidence that between the years of 1941 and 1991, trace elements (dietary components which are present only in minute quantities in our food but are essential for maintaining health, such as copper, iron, selenium, and zinc) **in conventionally grown fruit and vegetables had fallen by 76%.**

- Clearly, the debate concerning the nutritional values of conventional versus organically grown foods has some way to run and requires more scientific examination. Unfortunately, the FSA's view may be questioned as it is a government agency (remember BSE and the "healthy hamburger") while the SA, it could also be argued, is also biased in favour of organic food.

- **In addition, a recent 18 million euros, EU-funded project, showed that some organic vegetables (including spinach, cabbage, onions and tomatoes) and some fruit contain significantly higher levels of antioxidants (such as** vitamin C), flavones and beta-carotene than conventionally-farmed produce. Organic milk was also found to contain higher levels of antioxidants (such as vitamin E) and fatty acids, including omega 3. In addition, organic food was shown to contain lower levels of pesticides and heavy metals. The research was included in the EU "Quality Low Input Food (QLIF)" project and lead by Professor Carlo Leifert. The research involved over 31 institutes, companies and universities and took place over 5 years from March 2004 to April 2009. Most important was the fact that to achieve these results good agricultural practices were required and details of these were included in the reports (See, *Proceedings 3rd International Congress European Integrated Project Quality Low Input Food (QLIF), March 2-23, 2007 and www.qlif.org for leaflets and links).*

- Controversially, the FSA commissioned a report on organic versus conventionally produced food, published in 2009, which concluded that although differences exist between these foods, they were "not large enough to be of any public health significance" (*www.food.gov.uk/news*); **Dangour and colleagues, American Journal of Clinical Nutrition, Vol. 90, pages 680-685, 2009**). The FSA study seems to have ignored the Leifert reports (www.qlif.org) although details were being released of these in 2007.

WHO SHOULD THE POOR CONFUSED PUBLIC BELIEVE?

- The FSA report was not an original experimental study but a review of previously published work. The analyses included the results from studies published several decades ago probably before the best agricultural practices had been developed for organic food. The importance of such best practice was emphasized by Leifert and colleagues (**see:** _www.qlif.org_).

- In contrast to the FSA review, the Leifert QLIF reports were based on large scale, original experimental studies actually growing crops and rearing animals under optimised conventional and organic conditions.

- There has been widespread criticism of the FSA report for ignoring the QLIF reports although some "nutritionists" have accepted the findings blindly.

- It is tempting to accept the findings of the comprehensive QLIF study although criticism has been made of the lack of articles published from this study in peer-reviewed journals (**see, "The Organic Food Nutrition Wars" by Rosen,** _www.acsh.org/factsfears/newsid.1232/new s detail.asp_).

- However, a recent carefully designed study in 2010 on organic versus conventionally grown strawberries by Reganold and colleagues showed that organic strawberries not only contain higher antioxidant and phenolic compounds but also maintain a higher soil quality. See:

(*http://www.plosone.org/article/info%3Ad oi%2F10.1371%2Fjournal.pone.0012346*)

- In conclusion, there is some evidence of the nutritional superiority of organic produce although additional carefully designed studies are required.

- **A much more convincing argument for buying organic produce is because of the lower levels of pesticide residues in organic food. The FSA even supports this point of view.**

- There is now evidence to show that about 30% of our basic conventionally grown foods contain pesticide residues (see "*It's Your Life: End the Confusion from Inconsistent Health Advice*") but the effect of these residues on our health is debatable.

Two things, however, are clear:

1. **Our bodies contain cocktails of chemical pollutants** derived from pesticides in our food, from worming agents in our livestock, from chemicals contaminating the air we breathe and from products we use on our bodies or in our homes. These chemicals may interact in the body to compromise our health.

2. **We must not**, due to our fear of chemical contamination, **completely stop eating** vegetables and fruit grown conventionally. The benefits of the "five a day" policy must far outweigh the harm resulting from a diet free of vegetables and fruit. **(See Chapter 2 for details of which conventionally grown food to eat in order to reduce pesticides in your diet.)**

In addition, many people already believe that organic food:

1. Tastes better.

2. Has increased nutrient content.

3. Undoubtedly has reduced contamination with pesticides.

4. Has environmental benefits that preserve and encourage wildlife as a result of the use of fewer chemicals and from intercropping with food plants for beneficial insects to use for feeding and reproduction.

Other people point out, however, that with organic food:

1. Carbon emissions may be much higher as so much is imported.

2. Since yields are lower, more land is required for growing crops.

3. There are reports, for example in chickens, that organic food has higher contamination rates with harmful microorganisms.

From the above, there is limited evidence that organic fruit and vegetables contain higher levels of beneficial nutrients but much stronger is the proof of lower pesticide contamination in organic food. **Therefore it is logical to want to include some organic food in your diet.**

But it is not necessary or possible to replace all food items with organic produce.

Follow the simple rules below and you will not only reduce your pesticides intake, you will also take advantage of possibly higher levels of nutrients in some organic food, and you won't add excessively to your food bill each week.

OVERALL SUMMARY AND WHICH ORGANIC FOODS ARE MOST IMPORTANT TO BUY

- Generally, fruits and vegetables **once skinned, peeled, shelled or trimmed** will probably have pesticide contamination levels **reduced by as much as 85%** so organic is unnecessary.

- Most soft fruits (but not blueberries) have high pesticide levels so wash thoroughly (uncut). Organic grapes and strawberries, which often have high pesticide levels, are available but expensive.

- **Always buy organic apples/pears or peel before eating.**

- Many vegetables, including lettuce, celery, and tomatoes, as well as aubergines, peas with edible pods, green beans, and salad onions have high pesticide rates but buying organic is expensive or they are not available. Again, **washing thoroughly will help.** Locally grown produce may also have lower pesticide levels.

- **Always buy organic potatoes,** if you eat the skins, as these are usually available and cheap.

- Many other foods like meats and dairy produce have low pesticide levels so organic are unnecessary but trim off excess fat.

- **In conclusion, it is particularly recommended to buy organic and wholemeal bread, cereals, cereal bars, porridge, flour and potatoes.**

Also read "The Benefits of Organic Food Are Real":

http://endtheconfusion.wordpress.com

Additional Information

The information in this book has been derived from numerous scientific papers. If you are interested in finding out more then you can find details of these papers in my comprehensive book on health: "*It's Your Life: End the Confusion from Inconsistent Health Advice*". Here is why the book is unique:

- **It is for people of all different ages,** aiming to optimise health and fitness and maximise an active and independent lifestyle throughout life. It is not a part of the recent deluge of health and diet books or videos produced by B-class "celebrities" but has been written by a biomedical scientist of international repute.

- There are **high impact illustrations** emphasising important points in the text. For example, the cover illustrates the present-day frustration and confusion of the average consumer exposed to contradictory health and dietary advice.

- There are clear summaries of **basic facts for adopting a new health plan.** Thus, for the many people with busy lives who may hate reading health books Chapter 1 ("Food, The Basic Diet"), Chapter 9 ("Exercise, Basic Introduction") and other Chapters are designed for rapid reference, often to specific age-groups of people.

- You will also not find in most other books descriptions of how many aspects of **diet and exercise change at different times of life** (Chapter 1) as well as reasons for weight gain as we age and advice as how to avoid this (Chapter 2, "Help! What Am I Eating?").

- There are **extensive tables for rapid identification of foods containing high levels of calories, saturated fats, salt and sugar** (Chapter 2). Thus, information on over 300 different food groups can be extracted at a glance without the necessity of reading minute and confusing Supermarket Food Labels.

- You will also not find in most books **not only clearly tabulated facts** about "**The Good, The Bad And The Ugly Fats**", and "**Fibre**" but also appraisals of the Atkins and GI **"Fad Diets"** (Chapter 3).

- **Details of the rates of pesticide contamination of fruit, vegetables and other types of food** using easily interpreted tables are also given (Chapter 4). A summary table is also included, for attaching to the refrigerator door or notice board, to identify **the least chemically polluted foods.**

- **A list is given of which organic foods are the most important to buy** (Chapter 4) and an explanation why, in these financially challenged times, it is **unnecessary** to eat just organic foods.

- Details are also given **of the potential impact on food safety of Food Additives, Preservatives and Colourants** (Chapter 5) together with consideration of the **total chemical loading** of the body from all sources (Chapter 6, "The Cocktail Effect"). Possible interactions of chemicals accumulated from pesticides and additives in food, and from cosmetics and household sources, are also discussed, and advice is given on **reducing the uptake of chemicals from the environment**.

- You will also not find in other books an understanding of the **"Vitamin Dilemma" as "To Take Or Not To Take, That Is The Question"** (Chapters 7 and 8) facing most people following conflicting advice in the media. Clear scientific analysis of the latest research shows that people require different supplements at different stages in their lives. **Supplement recommendations are made for each stage from pregnancy to old age.**

- There is also **an understanding of the "To Gym Or Not To Gym-That Is The Question" dilemma faced by many people beginning to exercise for the first time** (Chapter 9). It introduces the basics and benefits of regular exercise, describes how to begin training in the gym, and provides an outline exercise programme (Chapter 10).

- **Details of "Alternative Types Of Exercise For Gym–Haters"** (Chapter 11), with different sports and activities are also described together with the calories used and **a table of the time taken with different sports to burn off highly calorific snacks.** Uniquely, the effects of each type of exercise are presented in terms of joint damage and cardiovascular function, and **advice on exercising at different ages** is also included.

- In summary, **"IT'S YOUR LIFE"**, presents the best advice available for optimising health and fitness in a manner designed to enlighten and engage the non-expert reader.

You can find details of where to buy the book here:

http://www.cranmorepublications.co.uk/6

In order to receive the latest health advice simply enter your email address here:

http://endtheconfusion.wordpress.com